Mary Kay Carson

Contents

Animal, Vegetable, or What?

Our world is full of fungi. A mushroom sprouts in the wet grass, a bracket fungus grows out of a log, and mold covers a forgotten orange. These are all kinds of fungi.

But what exactly is a fungus? Fungi are plantlike because they don't move around. But fungi can't make their own food as plants do. To live, fungi must eat more as animals do. But a mushroom isn't an animal either! Instead fungi make up their own kingdom, or separate group of life-forms.

Fungi are not plants because they lack **chlorophyll,** the chemical that makes plants green and helps them make their own food. Mysteriously, the cell walls of most fungi contain a tough substance called **chitin,** the stuff that covers and protects many insects!

When you say fungus, most people think of a mushroom. But a mushroom is only the fruiting body of a fungus, as an apple is only part of an apple tree. Most of the fungal body, the mushroom's **mycelium**, is hidden underground in the soil. The mycelium is a weblike network of thousands of tiny threads, called **hyphae**.

You can see the mold's mycelium growing on this orange.

The mycelium of this bracket fungus is hidden in the rotting tree trunk.

A fungus grows by spreading out its mycelium. Fruiting bodies such as mushrooms and puffballs sprout from the mycelium and push their way above ground. Their job is to make and release spores. Much as apple seeds sprout and grow into apple trees, spores sprout and grow into new fungi.

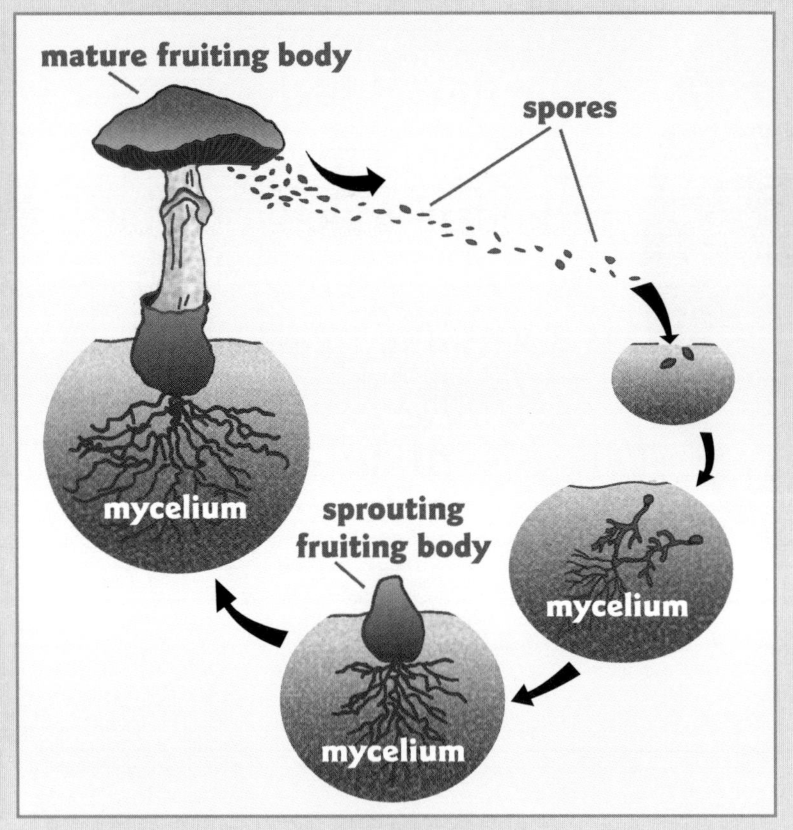

Quick Fact!

A single fungus can grow underground for hundreds of acres, making fungi the largest living things on Earth.

Fairy ring mushrooms sprout from the outer edge of an underground mycelium. As the mycelium grows outward, the center dies off and a circle is formed

But fungus spores are not the same as plant seeds; they are much smaller and simpler. Seeds include stored food for the new plant, but tiny spores don't. Instead, a fungus such as a puffball releases millions of simple, tiny, dustlike spores. Out of the millions of spores, only a few will land where there's food and grow into new fungi.

Different kinds of fungi release and spread their spores in different ways. The puffball releases puffs of spores into the air when drops of water fall on it. The stinkhorn fungus attracts flies to it with its rotten meat smell. The flies carry away spores on their legs.

Tour of the Kingdom

Scientists who study fungi are called **mycologists**. Mycologists find new species of fungi all the time, which they sort into fungal families within the large kingdom. Some fungi eat dead things. Some fungi digest the tissues of living plants and animals. Other fungi live in partnership with other living things.

This fungus is an insect **parasite.** Parasites eat living things, often harming and sometimes even killing them.

These fungi are growing on a rotting log. A fungus that eats dead things is a **saprophyte.**

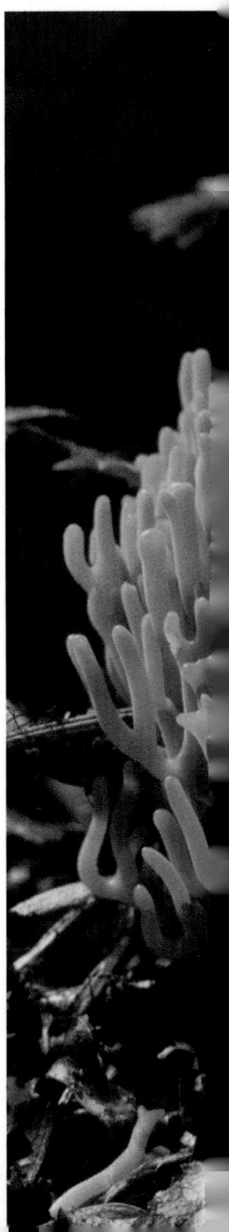

Symbiont fungi live in partnership with other living things. This lichen is actually a living partnership between a fungus and a tiny plant life-form called an alga.

Coral fungus

There are more than 150,000 different kinds of fungi and they come in a great variety of sizes, shapes, and colors. Check out these amazing members of Kingdom Fungi!

Bird's nest fungus

Nature's Recycler

The members of an ecosystem have different jobs. Plants are producers, producing or making their own food. Animals are consumers, consuming or eating foods such as plants or other animals. Fungi have a third job—they are decomposers.

These strawberries are being decomposed by the mold feeding on them. You digest food in your stomach after eating it, but a fungus digests its food first! This mold is oozing out **enzymes** to break down the strawberry tissues. Then the mold's hyphae can absorb the digested **tissue.**

Without fungi—and other decomposers such as bacteria, earthworms, and some insects—plants would run out of the nutrients they need to restart the cycle of life. Dead things such as leaves would quickly pile up and cover the earth. To see leaf mold at work, look for circle patterns.

Decomposers break down dead plants and animals into the basic chemical building blocks they're made of. These chemical nutrients go back into the soil where they are needed for more plants to grow.

Partners with Fungi

Many fungi live in partnership with a plant or an animal. Both members, called symbionts, benefit from the partnership. Most green plants have fungal symbionts living around their roots.

Fungi can change the minerals in soil into forms more easily used by plants. The fungi also deliver the minerals and water directly to the roots. In return, the fungi absorb some of the plants' stored food.

The partnership between a fungus and the roots of a plant is called **mycorrhiza,** which means "fungus-root." This is how mycorrhiza appears under a microscope. Scientists estimate that 80 to 90 percent of all green plants have fungal partners.

Together, as lichen, fungi and their partners can live in environments that they'd never be able to survive in as individuals. Lichens can thrive on bare rock, desert sand, or concrete.

Lichens are an amazing life-form made up of a fungus and an alga partner with chlorophyll that can manufacture its own food. The fungus eats some of the food that its partner makes. In turn, the fungus provides its partner with a moist, mineral-rich home protected from the weather.

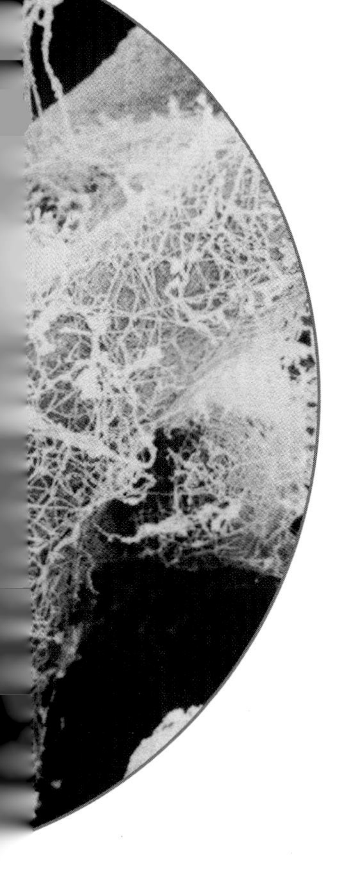

Americans call these lichens British soldiers and pixie cups. The two kinds of lichen are often found together.

Fungal Foes

Fungi are an important part of the ecosystem, but some fungi cause problems for people. Ringworm, athlete's foot, and nail fungus are all fungal diseases. Molds and other fungi, such as rust and corn smut, spoil food and rot grain, so scientists are constantly battling them.

Rust

Corn smut

Scientists at Work:
Keeping Fungal Pests Away

The beautiful cherry trees that grow in our nation's capital were the second batch of cherry trees sent from Japan. The first batch, a gift from the mayor of Tokyo, had to be destroyed.

Flora Patterson was the first woman mycologist at the U.S. Department of Agriculture when the first shipment of 2,000 trees arrived in 1910.

When inspectors found fungi and harmful insects on the trees, Patterson was afraid that the pests would spread to plants and crops nearby.

Patterson made the hard decision to burn the prized trees in a bonfire on the Mall.

Fungal Friends

While some fungi are harmful to humans, many are quite helpful. Edible mushrooms have been harvested from the wild and eaten by people for thousands of years. Today, new foods, medicines, and even vitamins are being made from fungi. Only experts can always tell wild edible mushrooms from deadly poisonous "toadstools." So never eat a fungus you find in the wild.

Oyster mushrooms are enjoyed for their delicate taste. They are gathered from the wild or farmed.

These edible mushrooms are farmed in complete darkness.

Scientists are discovering that fungi can be used to clean up toxic wastes in soil. While digesting and eating the dead things in contaminated soil, some fungi's enzymes also break down the hazardous chemicals. Fungi are important to humans and to the ecosystem.

Quick Fact!

As yeast grows inside bread dough, the fungus releases gases and creates air pockets that cause the bread to "rise."

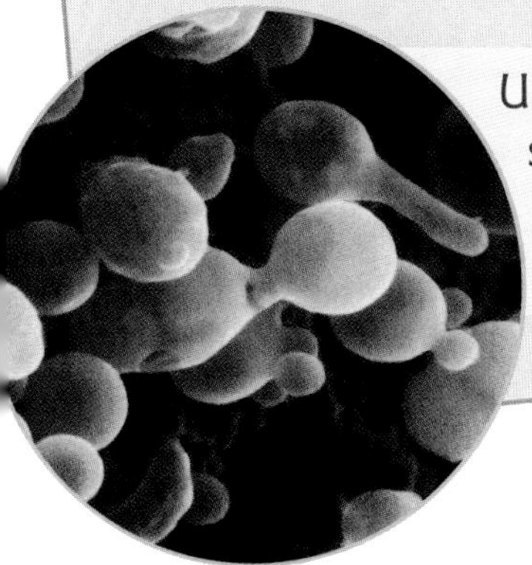

Under a microscope you can see how yeast just splits in two to reproduce. It's called budding.

Scientists at Work: A Wonder Drug From Mold

Alexander Fleming was a scientist who wanted to find a way to kill bacteria that are harmful to people. Returning to his lab after a vacation, he noticed mold in one of his dishes of bacteria. He took a closer look and noticed that the bacteria near the mold were dead. Fleming had discovered the bacteria-killing mold penicillium! Penicillin, a juice from the mold, was made into the world's first antibiotic. Antibiotics have saved millions of lives.

When Fleming was presented with the Nobel Prize for Medicine in 1945 he said, "Nature makes penicillin; I just found it."

Exploring for Fungi

Melissa Skrabal looks closely at the bark of a living tree. Excited, she thinks that she might be looking at the tracks left by a slime mold. She carefully takes a bark sample, places it in a bag, and lets it drop eight meters to the ground below. Melissa is part of a team of students and scientists that is attempting to document all living organisms in the Great Smoky Mountains National Park.

Since many fungi live high in the treetops or canopy, Melissa had to learn how to climb safely as high as 34 meters. Here she gathers a sample of lichen.

The Great Smoky Mountains National Park is one of the richest temperate biospheres in the world. That means that many different species of plants, animals, and fungi live here.

As Melissa continues to climb, she hears a lot of hooting and celebrating from the ground team below. When the team looked at her sample with a hand lens they saw a glittering gold and pink organism. They knew it was rare. It turned out to be a new species of slime mold.

This is how Melissa's slime mold looks under a microscope.

Quick Fact!

Until recently, slime molds were included in the Kingdom Fungi. But slime molds lack chitin in their cells and they don't digest their food outside their bodies. They actually swim, crawl, or ooze to surround their food and then digest it. So scientists tossed slime molds out of Kingdom Fungi.

In Hawaii, 15-year-old Kolea packed his backpack with equipment, pulled on his jeans and a pair of rubber boots, and headed for the rain forest. His mission was to find fungi. The forest was the Pu‘u Maka‘ala Natural Area Reserve. Because they are so hard to find in the dense undergrowth, many of Hawaii's fungi have not yet been identified. Kolea found 16 different species and won a Young Naturalist Award in 2002 for his work.

Kolea's field notes show and describe each of the different species of fungi he found that day. Ten of them were rare. "Who knows what important antibiotics and pharmaceuticals they contain?"

Of the new species of fungi found, none were exotic. That means that none were from faraway places. "The lack of exotic fungi," writes Kolea, "shows that the forest is extremely pristine." A pristine forest is one that has not been changed by people.

Investigate Mold

You need two disposable jars with lids, heavy tape, water, a slice of bread, grapes, a permanent marker, and a hand lens.

1. Why do foods mold? *Record* what you know about mold.

2. Cut the bread into pieces. Divide both the bread and the grapes between the two jars. Label one jar A and the other B.

3. Sprinkle a little water into jar A, attach the lid, use the tape to seal it shut, and place it away from sunlight.

4. As a variable, leave jar B open for two hours, then repeat step 3. *Predict* what will happen in each of the jars.

5. Observe your jars using a hand lens every day for ten days.

Don't open them!

Record your *observations* every day, along with the date and a drawing of each jar's contents.

When finished, place the unopened jars in the trash.

6. What can you *conclude* about how mold grows?

Glossary

chitin (KYE-ten) a tough substance that forms part of the protective outer covering of some insects and the cell walls of fungi

chlorophyll (KLOR-uh-fil) the green coloring in plants necessary for them to make food from air and water through photosynthesis

enzyme (EN-zym) a substance made by living cells that produces chemical changes

hypha (HIGH-fuh), plural *hyphae* (HY-fee), one of the thin threads that make up the body of a fungus

mycelium (my-SEE-lee-um), plural *mycelia* (my-SEE-lee-uh), a network of hyphae that make up the body of a fungus

mycologist (my-KOL-uh-jist) a scientist who studies fungi

mycorrhiza (MY-kuh-RYE-zuh) the partnership between a soil fungus and a living plant root

parasite (PAIR-uh-syt) a life-form, such as a fungus, that absorbs nutrients from living things, often harming them

saprophyte (SA-proh-fite) a life-form that absorbs nutrients from dead things

symbiont (SIM-bee-ont) a life-form that lives in partnership with another life-form

tissue (TIH-shoo) a mass of similar cells that together form part of an organism